# Workbook of the Basics of Probability:

## A Maths Self-Assessment Guide

By

Kingsley Augustine

# Table of Contents


NOTE ON USE OF THIS WORKBOOK ........ 3

CHAPTER 1 THE BASIC THEORY OF PROBABILITY ........ 4

CHAPTER 2 PROBABILITY ON SIMPLE EVENTS ........ 5

CHAPTER 3 PROBABILITY ON PACK OF PLAYING CARDS ........ 16

CHAPTER 4 PROBABILITY ON TOSSING OF COINS ........ 43

CHAPTER 5 PROBABILITY ON THROWING OF DICE ........ 52

CHAPTER 6 MISCELLANEOUS PROBLEMS ON PROBABILITY ........ 66

# NOTE ON USE OF THIS WORKBOOK

In order to ensure the direct use of this workbook, blank spaces are provided below each question. A blank space should be used by learners to solve the question that is directly above it. Answers have been provided at the end of each chapter for students to check if what they have solved is correct. I strongly advice students to solve questions in this workbook by using pencils so that they can always erase what they have written and re-solve again if what they solved was not correct.

# CHAPTER 1
# THE BASIC THEORY OF PROBABILITY

Probability is the likelihood of an event happening. Mathematically probability is given by:

$$\text{Probability} = \frac{\text{number of required outcome}}{\text{number of total or possible outcome}}$$

If the probability of an event happening is x, then the probability that it will not happen will be given by:

$$1 - x$$

Probability must lie between the values of 0 and 1. If an event cannot happen, then its probably is 0. If an event is certain to happen, then its probability is 1.

**Mutually Exclusive Events**

When there is no member/element common between two or more similar events, then we say they are mutually exclusive events. For example the event of odd numbers or even numbers are mutually exclusive. They are disjoint sets.

**Addition Law of Probability**

If two events are mutually exclusive, then the probability of one or the other happening is the sum of their individual probabilities.

**Independent Events**

When a die is thrown, and a coin is tossed, these two events have no effect on each other. Such events are called independent events

## Product law of probability

If two events are independent, then the probability of both events happening is the product is the product (multiplication) of their individual probabilities.

# CHAPTER 2
# PROBABILITY ON SIMPLE EVENTS

**Examples**

1. The table below give the number of students in each age group in a class.

| Age (Years) | 12 | 13 | 14 | 15 | 16 | 17 |
|---|---|---|---|---|---|---|
| number of students | 6 | 3 | 10 | 4 | 2 | 5 |

If a student is chosen at random, find the probability that the student is:
(a) 13 years old
(b) 15 years old or less
(c) at least 16 years old
(d) most 13 years old
(e) not 17 years old

**Solution**

(a) Pr. (13 years old) $= \frac{\text{Number of students who are 13 years old}}{\text{Total number of students}}$

$= \frac{3}{30}$

$= \frac{1}{10}$ (when $\frac{3}{30}$ is express in its lowest term, it gives $\frac{1}{10}$)

(b) Pr. (15 years or less) $= \frac{\text{Students who are 15 years and below}}{\text{Total number of students}}$

$= \frac{4+10+3+6}{30}$

$= \frac{23}{30}$

(c) Pr. (At least 16 years old) $= \frac{\text{Students who are 16 years and above}}{\text{Total number of students}}$

$= \frac{2+5}{30}$

$= \frac{7}{30}$

(d) Pr. (At most 13 years) $= \frac{\text{Students who are 13 years and below}}{\text{Total number of students}}$

$= \frac{3+6}{30}$

$= \frac{9}{30}$

$= \frac{3}{10}$ (When expressed in its lowest term)

(e) Pr. (17 years old) = $\frac{\text{number of students who are 17 years old}}{\text{total number of students}}$

$= \frac{5}{30}$

$= \frac{1}{6}$

Therefore, Pr. (Not 17 years old) = 1 - Pr. (17 years old)

$= 1 - \frac{1}{6}$

$= \frac{5}{6}$

2. The probability that a seed will germinate is $\frac{2}{5}$. What is the probability that it will not germinate?

**Solution**

Pr. (It will germinate) $= \frac{2}{5}$

Pr. (It will not germinate) $= 1 - \frac{3}{5}$

$= \frac{2}{5}$

3. A letter is chosen at random from the alphabet. Find the probability that it is one of the letters of the word: PROBABILITY.

**Solution**

In this case a letter should not be counted more than once. Avoiding repetition, the word can now be written as:

PROBALITY (i.e. 9 letters). Note that there are 26 letters of the alphabet.

Therefore, Pr. (one letter from PROBABILITY) = $\frac{9}{26}$

4. A number is chosen at random between 1 and 16, both inclusive. What is the probability that it is:
(a) even
(b) prime
(c) odd or prime
(d) divisible by 4
(e) a perfect square or a perfect cube

**Solution**

(a) Total numbers in all from 1 to 16 = 16

The even numbers are 2, 4, 6, 8, 10, 12, 14, 16

Therefore the number of even numbers is 8

Hence Pr. (even number selected) = $\frac{\text{Number of even numbers}}{\text{Total numbers in all}}$

$= \frac{8}{16}$

$= \frac{1}{2}$

(b) The prime numbers are 2, 3, 5, 7, 11, 13

Therefore the number of prime numbers is 6

Hence Pr. (prime number selected) = $\frac{\text{Number of prime numbers}}{\text{Total numbers in all}}$

$= \frac{6}{16}$

$= \frac{3}{8}$

(c) The odd numbers are 1, 3, 5, 7, 9, 11, 13, 15

The prime numbers are 1, 3, 5, 7, 11, 13

Since OR in probability means addition, then we add all the odd and prime numbers together, but we must not count any number twice. This gives 1, 3, 5, 7, 9, 11, 13, 15, which is a total of 8 numbers.

Hence Pr. (odd or prime number selected) = $\frac{\text{Number of odd and even numbers}}{\text{Total numbers in all}}$

$= \frac{8}{16}$

$= \frac{1}{8}$

(d) The numbers divisible by 4 are 4, 8, 12, 16

This gives a total of 4 numbers

Hence Pr. (a number divisible by 4) = $\frac{\text{The four numbers divisible by 4}}{\text{Total numbers in all}}$

$= \frac{4}{16}$

$= \frac{1}{4}$

(e) The perfect square numbers are 1, 4, 9, 16

The perfect cube numbers are 1, 8

Since OR in probability means addition, then we add all the set of values above without counting any number twice. This gives 1, 4, 8, 9, 15, which is a total of 5 numbers.

Hence Pr. (perfect square or perfect cube selected) = $\frac{15}{16}$

5. A letter is chosen at random from the alphabet. Find the probability that it is:
(a) T
(b) E or P
(c) not B or G
(d) either D, J, N, U, W or Y
(e) one of the letters of the word REJECTED

**Solution**
(a) There are 26 letters of the alphabet, out of which there is 1 T.

Therefore, Pr. (T) = $\frac{\text{Number of Ts}}{\text{Total numbers of alphabets}}$

$$= \frac{1}{26}$$

(b) Pr. (E or P) = $\frac{\text{Number of Es and Ps}}{\text{Total numbers of alphabets}}$

$$= \frac{6}{26}$$

$$= \frac{1}{13}$$

(c) Pr. (B or G) = $\frac{2}{26} = \frac{1}{13}$

Therefore, Pr. (not B or G) = 1 - Pr. (B or G)

$$= 1 - (\frac{6}{13})$$

$$= \frac{12}{13}$$

(d) The letters D, J, N, U, W and Y makes a total of 6 letters.

Pr. (D, J, N, U, W or Y) = $\frac{6}{26}$

$$= \frac{3}{13}$$

(e) Writing the letters of the word REJECTED without repeating a letter gives REJCTD. This gives a total of 6 letters

Therefore Pr. (one of the letters of REJECTED) = $\frac{6}{26}$

$$= \frac{3}{13}$$

6. A letter is selected at random from the word PROBABILITY. What is the probability of selecting the letter B.

**Solution**

In this case the total letters of the word PROBABILITY gives 11. The repeated letters should be counted more than once since this is not a case of letter from the alphabet. In the 26 alphabet each letter appears once, that is why they are counted once. But in PROBABILITY (or other words that might be given) some letters appear more than once, hence they should be counted as many times as they appear.

In PROBABILITY, B appears 2 times.

Therefore, Pr. (selecting B) = $\frac{2}{11}$

**Practice Questions**

1. The table below give the number of students in each mark group in a class.

| Mark | 5 | 6 | 7 | 8 | 9 | 10 |
|---|---|---|---|---|---|---|
| Number of students | 3 | 6 | 2 | 4 | 1 | 4 |

If a student is chosen at random, find the probability that the student scored:
(a) 7 marks
(b) 6 marks or less
(c) at least 9 marks
(d) at most 8 marks
(e) 5 or 8 maks

**Solution**

(a)

(b)

(c)

(d)

(e)

2. The probability that a seed will germinate is $\frac{3}{4}$. What is the probability that it will not germinate?

**Solution**

3. A letter is chosen at random from the alphabet. Find the probability that it is one of the letters of the word: MATHEMATICS.

**Solution**

4. The probability that a man wins an election is $\frac{3}{5}$. What is the probability that he does not win.

**Solution**

5. Out of every 10 bulbs, 2 do not last long. What is the probability that a bulb will last long when lit?

**Solution**

6. In family, the number of males is 3, while the number of females is 2. Find the probability that another child born into the family is:
(a) a male child
(b) a female child

**Solution**

(a)

(b)

7. A survey shows that 44% of all women take size 7 shoes. What is the probability that a mother of two takes size 7 shoes?

**Solution**

8. In a secondary school, 30 out of every 100 students are at least 160cm tall. What is the probability that a student chosen at random from the school is less than 160cm tall?

**Solution**

9. A number is chosen at random between 1 and 20, both inclusive. What is the probability that it is:
(a) prime
(b) odd

(c) even or prime
(d) divisible by 3
(e) a number less than 10 or a perfect cube

**Solution**

(a)

(b)

(c)

(d)

(e)

10. A letter is chosen at random from the alphabet. Find the probability that it is:
(a) F
(b) M or Q or Y
(c) in the word COME
(d) either in the word BUT or in REMOVE
(e) one of the letters of the word SURPRISED

**Solution**

(a)

(b)

(c)

(d)

(e)

11. A letter is selected at random from the word RESPIRATION. What is the probability of selecting the letter I.

**Solution**

## Answers to Chapter 2

1. (a) $\frac{1}{10}$ (b) $\frac{9}{20}$ (c) $\frac{1}{4}$ (d) $\frac{3}{4}$ (e) $\frac{7}{20}$

2. $\frac{1}{4}$ 3. $\frac{4}{13}$ 4. $\frac{2}{5}$ 5. $\frac{4}{5}$

6. (a) $\frac{3}{5}$ (b) $\frac{2}{5}$

7. $\frac{11}{25}$ 8. $\frac{7}{10}$

9. (a) $\frac{2}{5}$ (b) $\frac{1}{2}$ (c) $\frac{9}{10}$ (d) $\frac{3}{10}$ (e) $\frac{9}{20}$

10. (a) $\frac{1}{26}$ (b) $\frac{3}{26}$ (c) $\frac{2}{13}$ (d) $\frac{4}{13}$ (e) $\frac{4}{13}$

11. $\frac{2}{11}$

# CHAPTER 3
# PROBABILITY ON PACK OF PLAYING CARDS

A pack of playing cards contains 52 cards of 4 types. There are 13 clubs, 13 diamonds, 13 hearts and 13 spades. Each of the set of 13 cards contains Ace (A), 2, 3, 4, 5, 6, 7, 8, 9, 10, Jack (J), Queen (Q), and King (K). This means that out of the 52 cards, each card is four in number, i.e. Aces are 4 in number, 1s are 4 in number, 2s are 4 in number, 3s are 4 in number, 4s are 4 in number, 5s are 4 in number, 6s are 4 in number, 7s are 4 in number, 8s are 4 in number, 9s are 4 in number, 10s are 4 in number, Jacks are 4 in number, Queens are 4 in number, and Kings are 4 in number. Clubs and spades are black, diamonds and hearts are red. This means that there are 26 black cards and 26 red cards. This also means that out of the 4 Aces cards, 2 are black and 2 are red. Out of the four cards that are 1s, two are black and two are red, out of the four cards that are 2, two are black and two are red, and so on.

**Examples**

1. A card is picked at random from a pack of playing cards. Find the probability of picking a spade.

**Solution**

There are 13 spades in a pack of playing cards.

Therefore, Pr. (picking a spade) = $\frac{\text{Number of Spades}}{\text{Total numbers of cards}}$

$= \frac{13}{52}$

$= \frac{1}{4}$ (In its lowest term)

2. A card is picked at random from a pack of playing cards. Find the probability of picking a red card.

**Solution**

There are 26 red cards in a pack of playing cards.

Therefore, Pr. (picking a red card) = $\frac{\text{Number of red cards}}{\text{Total numbers of cards}}$

$= \frac{26}{52}$

$$= \frac{1}{2} \quad \text{(In its lowest term)}$$

3. A card is picked at random from a pack of playing cards. Find the probability of picking a red 5.

**Solution**

There are 2 red 5 cards in a pack of playing cards.

Therefore, Pr. (picking a red 5) = $\frac{\text{Number of red 5}}{\text{Total numbers of cards}}$

$$= \frac{2}{52}$$

$$= \frac{1}{26} \quad \text{(In its lowest term)}$$

4. A card is picked at random from a pack of playing cards. Find the probability of picking a 3.

**Solution**

There are 4 cards that are 3 in a pack of playing cards.

Therefore, Pr. (picking a 3) = $\frac{\text{Number of cards that are 3}}{\text{Total numbers of cards}}$

$$= \frac{4}{52}$$

$$= \frac{1}{13} \quad \text{(In its lowest term)}$$

5. A card is picked at random from a pack of playing cards. Find the probability of picking

(a) a black or red card

(b) a 2 or a 5

(c) either a heart or the king of spades

(d) a club or a red Queen

(e) a diamond or a 9

(f) a 6 or a black card

**Solution**

(a) There are 26 black cards and 26 red card

Since or in probability means plus, then we have to add the numbers. This gives a total of: 26 + 26 = 52

Therefore, Pr. (picking a black or red card) = $\frac{\text{Number of black and red cards}}{\text{Total numbers of cards}}$

$= \frac{52}{52}$

= 1

(b) There are 4 cards that are 2, and 4 cards that are 5. This gives a total of 8 cards.

Therefore, Pr. (picking a 2 or a 5) = $\frac{8}{52}$

$= \frac{2}{13}$

(c) There are 13 cards that are Hearts, and 1 king that is a spade. This gives a total of 14 cards.

Therefore, Pr. (picking either a heart or the king of spades) = $\frac{14}{52}$

$= \frac{7}{26}$

(d) There are 13 cards that are club, and 2 cards that are red Queen, (i.e. the Queen of hearts and the queen of diamond). This gives a total of 15 cards.

Therefore, Pr. (picking a club or a red Queen) = $\frac{15}{52}$

(e) There are 13 cards that are diamonds, and 4 cards that are 9. But one of the 9 is in diamond and has already been counted among the 13 diamonds. So it must not be counted twice. Hence we count the other three 9 (each from clubs, hearts and spades). This will give a total of 16 (13 + 3) cards.

Therefore, Pr. (picking a diamond or a 9) = $\frac{16}{52}$

$= \frac{4}{13}$

(f) There are 4 cards that are 6, and 26 cards that are black. But two of the 26 black cards are among the four cards that are 6, and these two black 6 cards have already been counted among the 26 black cards. So they must not be counted twice. Hence we count the other two 6 cards that are red. This will give a total of 28 (26 + 2) cards.

Therefore, Pr. (picking a 6 or a back card) = $\frac{28}{52}$

$= \frac{7}{13}$

6. A card is picked at random from a pack of playing cards and then replaced. A second card is picked. What is the probability of picking:

(a) a 3 and a 10

(b) a queen and an ace

(c) two kings

(d) two red cards

(e) two cards of different colour

(f) two cards of the same colours

**Solution**

In probability problems, when two items are selected, it is important to logically analyse the situation when solving the problem. This will help you to know if addition (use of OR) is involved or multiplication (use of AND) is involved. For example, for a queen and a king to be selected, it simply means that, either the queen is selected first and then the king, or the king is selected first and then the queen. When this logical analysis is understood, then most questions in probability become easy to solve.

(a) There are four cards that are 3, and four cards that are 10

Therefore, Pr. (picking a 3) = $\frac{4}{52}$

$= \frac{1}{13}$

Similarly, Pr. (picking a 10) = $\frac{4}{52}$

$= \frac{1}{13}$

Recall that "and" in probability means multiplication.

The probability of picking a 3 and a 10 means that:

Either the first is a 3 AND the second is a 10, OR the first is a 10 AND the second is a 3.

This can be calculated by putting x in place of AND and + in place of OR in the above statement as follows:

Pr. (picking a 3) x Pr. (picking a 10) + Pr. (picking a 10) x Pr. (picking a 3)

$$= (\frac{1}{13} \text{ x } \frac{1}{13}) + (\frac{1}{13} \text{ x } \frac{1}{13})$$

$$= \frac{1}{169} + \frac{1}{169}$$

$$= \frac{2}{169}$$

Therefore, Pr. (picking a 3 and a 10) = $\frac{2}{169}$

(b) There are 4 cards that are queen, and 4 cards that are ace

Therefore, Pr. (picking a queen) = $\frac{4}{52}$

$$= \frac{1}{13}$$

Similarly, Pr. (picking an ace) = $\frac{4}{52}$

$$= \frac{1}{13}$$

The probability of picking a queen and an ace means that:

Either you first pick a queen AND then an ace, OR you first pick an ace AND then a queen.

This can be calculated by putting x in place of AND and + in place of OR in the above statement as follows:

Pr. (picking a queen) x Pr. (picking an ace) + Pr. (picking an ace) x Pr. (picking a queen)

$$= (\frac{1}{13} \text{ x } \frac{1}{13}) + (\frac{1}{13} \text{ x } \frac{1}{13})$$

$$= \frac{1}{169} + \frac{1}{169}$$

$$= \frac{1}{169}$$

Therefore, Pr. (picking a queen and an ace) = $\frac{2}{169}$

(c) There are four cards that are King

Therefore, Pr. (picking a king) = $\frac{4}{52}$

$= \frac{1}{13}$

The probability of picking two kings means that:

The first is a king AND the second is a king

= Pr. (picking a king) x Pr. (picking a king)

$= \frac{1}{13} \times \frac{1}{13}$

$= \frac{1}{169}$

Therefore, Pr. (picking two kings) = $\frac{1}{169}$

(d) There are 26 cards that are red

Therefore, Pr. (picking a red card) = $\frac{26}{52}$

$= \frac{1}{2}$

The probability of picking two red cards means that:

The first is a red card AND the second is a red card

= Pr. (picking a red card) x Pr. (picking a red card)

$= \frac{1}{2} \times \frac{1}{2}$

$= \frac{1}{4}$

Therefore, Pr. (picking two red cards) = $\frac{1}{4}$

(e) There are two colours of cards, red and black.

Therefore, Pr. (picking a red card) = $\frac{1}{2}$ (i.e from $\frac{26}{52}$ since there are 26 red cards)

Similarly, Pr. (picking a black card) = $\frac{1}{2}$ (i.e from $\frac{26}{52}$ since there are also 26 black cards)

The probability of picking two cards of different colours means that:

Either the first is a black card AND the second is a red card, OR the first is a red card AND the second is a black card.

This can be calculated by putting x in place of AND and + in place of OR in the above statement as follows:

Pr. (picking a black card) x Pr. (picking a red card) + Pr. (picking a red card) x Pr. (picking a black card)

$$= (\frac{1}{2} \times \frac{1}{2}) + (\frac{1}{2} \times \frac{1}{2})$$

$$= \frac{1}{4} + \frac{1}{4}$$

$$= \frac{2}{4}$$

$$= \frac{1}{2}$$

Therefore, Pr. (picking two cards of different colours) = $\frac{1}{2}$

(f) Pr. (picking two cards of the same colours) = 1 - Pr. (picking two cards of different colours)

$$= 1 - \frac{1}{2}$$

$$= \frac{1}{2}$$

Note that this can also be solved by using the logical process which is:

Either the first is red AND the second is red OR the first is black AND the second is black. This will also give $\frac{1}{2}$

7. Two cards are picked at random one after the other without replacement from a pack of playing cards. What is the probability of picking:

(a) a 5 and a 7

(b) a king and a jack

(c) two aces

(d) two diamond cards
(e) two black cards
(f) a red and a black card
(g) two cards of the same colours

**Solution**

This problem involves picking a card without replacement. This means that when one card is picked out, the total number of cards remaining in the pack become reduced to 51. That number of that particular type of card also reduces by 1.

(a) There are four cards that are 5. There are also four cards that are 7.

Hence the probability of picking a 5 and a 7 means that:

Either first picking a 5 AND then a 7, OR first picking a 7 AND then a 5.

Now, let us calculate each of the probabilities as follows:

Pr. (first card is a 5) = $\frac{4}{52}$ (There are four cards that are 5)

= $\frac{1}{13}$ (In its lowest term)

We now have 51 cards left in the pack.

Therefore, Pr. (second card is a 7) = $\frac{4}{51}$ (There are four cards that are 7, and a total of 51 cards remaining in the pack)

Or,

Pr. (first card is a 7) = $\frac{4}{52}$ (There are four cards that are 7)

= $\frac{1}{13}$ (In its lowest term)

We now have 51 cards left in the pack.

Therefore, Pr. (second card is a 5) = $\frac{4}{51}$ (There are four cards that are 5, and a total of 51 cards remaining in the pack)

Hence the probability of picking a 5 and a 7 means that:

Either first picking a 5 AND then a 7, OR first picking a 7 AND then a 5. Which is computed as:

Pr. (picking a 5 and a 7) = Pr. (first card is a 5) x Pr. (second card is a 7) + Pr. (first card is a 7) x Pr. (second card is a 5)

$$= (\frac{1}{13} \times \frac{4}{51}) + (\frac{1}{13} \times \frac{4}{51})$$

$$= \frac{4}{663} + \frac{4}{663}$$

$$= \frac{8}{663}$$

(b) There are four cards that are kings. There are also four cards that are jacks.

Now, let us calculate each of the probabilities as follows:

Pr. (first card is a king) = $\frac{4}{52}$ (There are four cards that are kings)

$= \frac{1}{13}$ (In its lowest term)

We now have 51 cards left in the pack.

Therefore, Pr. (second card is a jack) = $\frac{4}{51}$ (There are four cards that are jack, and a total of 51 cards remaining in the pack)

Or,

Pr. (first card is a jack) = $\frac{4}{52}$ (There are four cards that are jack)

$= \frac{1}{13}$ (In its lowest term)

We now have 51 cards left in the pack.

Therefore, Pr. (second card is a king) = $\frac{4}{51}$ (There are four cards that are king, and a total of 51 cards remaining in the pack)

Hence the probability of picking a king and a jack means that:

Either first picking a king AND then a jack, OR first picking a jack AND then a king. This is computed as:

Pr. (picking a king and a queen) = Pr. (first card is a king) x Pr. (second card is a jack) + Pr. (first card is a jack) x Pr. (second card is a king)

$$= (\frac{1}{13} \times \frac{4}{51}) + (\frac{1}{13} \times \frac{4}{51})$$

$$= \frac{4}{663} + \frac{4}{663}$$

$$= \frac{8}{663}$$

(c) There are 4 cards that are aces.

Hence the probability of picking two aces means that:

The first is an ace, and the second is an ace.

Now, let us calculate each of the probabilities as follows:

Pr. (first card is an ace) = $\frac{4}{52}$ (There are 4 cards that are aces)

$= \frac{1}{13}$ (In its lowest term)

We now have 3 aces left in the pack, and a total of 51 cards left in the pack.

Therefore, Pr. (second card is an ace) = $\frac{3}{51}$

Hence the probability of picking two aces is given by:

Pr. (picking two aces) = Pr. (first card is an ace) x Pr. (second card is an ace)

$$= \frac{1}{13} \times \frac{3}{51}$$

$$= \frac{3}{663}$$

(d) There are 13 cards that are diamonds.

Hence the probability of picking two diamonds means that:

The first is a diamond, and the second is a diamond.

Now, let us calculate each of the probabilities as follows:

Pr. (first card is a diamond) = $\frac{13}{52}$ (There are 13 cards that are diamonds)

$= \frac{1}{4}$ (In its lowest term)

We now have 12 diamonds left in the pack, and a total of 51 cards left in the pack.

Therefore, Pr. (second card is a diamond) = $\frac{1}{13}$

$= \frac{4}{17}$ (In its lowest term)

Hence the probability of picking two diamonds is given by:

Pr. (picking two diamonds) = Pr. (first card is a diamond) x Pr. (second card is a diamond)

$= \frac{1}{4} \times \frac{4}{17}$

$= \frac{4}{68}$

$= \frac{1}{17}$ (In its lowest term)

(e) There are 26 black cards.

Hence the probability of picking two black cards means that:

The first is a black card, and the second is a black card.

Now, let us calculate each of the probabilities as follows:

Pr. (first card is a black card) = $\frac{26}{52}$

$= \frac{1}{2}$ (In its lowest term)

We now have 25 black cards left in the pack, and a total of 51 cards left in the pack.

Therefore, Pr. (second card is a black card) = $\frac{25}{51}$

Hence the probability of picking two black cards is given by:

Pr. (picking two black cards) = Pr. (first card is a black card) x Pr. (second card is a black card)

$$= \frac{1}{2} \times \frac{25}{51}$$

$$= \frac{25}{102}$$

(f) The logical explanation for this situation is that:

Either the first card is red AND the second is black OR the first card is black and the second is red.

There are 26 red cards and also 26 black cards.

Now, let us calculate each of the probabilities as follows:

Pr. (first card is a red card) = $\frac{26}{52}$

$= \frac{1}{2}$ (In its lowest term)

We now have 51 cards left in the pack.

Therefore, Pr. (second card is a black card) = $\frac{26}{51}$ (There are 26 black cards, and a total of 51 cards remaining in the pack)

Or,

Pr. (first card is a black card) = $\frac{26}{52}$

$= \frac{1}{2}$ (In its lowest term)

We now have 51 cards left in the pack.

Therefore, Pr. (second card is a red card) = $\frac{26}{51}$ (There are 26 red cards, and a total of 51 cards remaining in the pack)

Hence the probability of picking a red card and a black card means that:

Either first picking a red card AND then a black card, OR first picking a black card AND then a red card. This is computed as:

Pr. (picking a red and black cards) = Pr. (first card is a red card) x Pr. (second card is a black card) + Pr. (first card is a black card) x Pr. (second card is a red card)

$$= (\frac{1}{2} \text{x} \frac{26}{51}) + (\frac{1}{2} \text{x} \frac{26}{51})$$

$$= \frac{26}{102} + \frac{26}{102}$$

$$= \frac{52}{102}$$

$$= \frac{26}{51}$$

(g) The logical explanation for this situation is that:

Either the first card is red AND the second is red OR the first card is black and the second is black.

There are 26 red cards and also 26 black cards.

Now, let us calculate each of the probabilities as follows:

Pr. (first card is a red card) = $\frac{26}{52}$

$= \frac{1}{2}$ (In its lowest term)

We now have 25 red cards left and a total of 51 cards left in the pack.

Therefore, Pr. (second card is a red card) = $\frac{25}{51}$

Or,

Pr. (first card is a black card) = $\frac{26}{102}$

$= \frac{1}{2}$ (In its lowest term)

We now have 25 black cards left and a total of 51 cards left in the pack.

Therefore, Pr. (second card is a black card) = $\frac{25}{51}$

Hence the probability of picking two cards of the same colour means that:

Either picking a red card AND then another red card, OR picking a black card AND then another black card. Which is computed as:

Pr. (picking two cards of the same colour) = Pr. (first card is a red card) x Pr. (second card is a red card) + Pr. (first card is a black card) x Pr. (second card is a black card)

$$= (\frac{1}{2} \times \frac{25}{51}) + (\frac{1}{2} \times \frac{25}{51})$$

$$= \frac{25}{102} + \frac{25}{102}$$

$$= \frac{50}{102}$$

$$= \frac{25}{51}$$

Alternatively, this question can also be solved as follows:

Recall that question (f) above gives the probability of picking a red and a black card. This also means the probability of picking two cards of different colours.

Hence the probability of picking two cards of different colours as given in (f) above = $\frac{26}{51}$

Therefore, Pr. (picking two cards of the same colour) = 1 - Pr. (picking two cards of different colours) (Note that they are opposite statements)

$$= 1 - \frac{26}{51}$$

$$= \frac{51-26}{51}$$

$$= \frac{25}{51} \quad \text{(As obtained before)}$$

8. If three cards are chosen from a pack of playing cards without replacement, what is the probability of getting:
(a) at least two diamonds
(b) at most one diamond?

**Solution**

Now, in order to write out the outcomes, let us use the letter D to represent a diamond and letter N to represent not a diamond.

Hence the outcomes are written as follows:

(DDD), (DDN), (DND), (DNN), (NDD), (NDN), (NND), (NNN)

(a) In order to determine the probability of getting at least two diamonds, we need to compute the probabilities of the brackets that contain at least 2 diamonds. They are, (DDD), (DDN), (DND), and (NDD).

Hence the probability of getting at least two diamonds = (DDD) or (DDN) or (DND) or (NDD)

Now, let us compute each of the probabilities.

There are 13 diamonds in a pack of cards, and there are 39 cards that are not diamonds. Note that this is a case of without replacement, which means that after each selection, both the total number of cards left and the number of the particular card picked, are reduced by 1. Hence:

(DDD) = Pr. (first card is a diamond) x Pr. (second card is a diamond) x Pr. (third card is a diamond)

$= \frac{13}{52} \text{X} \frac{12}{51} \text{X} \frac{11}{50}$ (Note that the number of diamond and the total number of card left, keep reducing by 1 after each selection)

$= \frac{1}{4} \text{X} \frac{12}{51} \text{X} \frac{11}{50}$

$= \frac{132}{10200}$

$= \frac{11}{850}$ (In its lowest term, after equal division by 12)

(DDN) = Pr. (first card is a diamond) x Pr. (second card is a diamond) x Pr. (third card is not a diamond)

$= \frac{13}{52} \text{X} \frac{12}{51} \text{X} \frac{39}{50}$ (Note that there are 39 cards that are not diamond)

$= \frac{1}{4} \text{X} \frac{4}{17} \text{X} \frac{39}{50}$

$= \frac{156}{3400}$

$= \frac{39}{850}$ (After equal division by 4)

(DND) = Pr. (first card is a diamond) x Pr. (second card is not a diamond) x Pr. (third card is a diamond)

$= \frac{13}{52} \text{X} \frac{39}{51} \text{X} \frac{12}{50}$

$$= \frac{1}{4} \times \frac{13}{17} \times \frac{6}{25}$$

$$= \frac{78}{1700}$$

$$= \frac{39}{850}$$

(NDD) = Pr. (first card is not a diamond) x Pr. (second card is a diamond) x Pr. (third card is a diamond)

$$= \frac{39}{52} \times \frac{13}{51} \times \frac{12}{50}$$

$$= \frac{3}{4} \times \frac{13}{51} \times \frac{6}{25}$$

$$= \frac{234}{5100}$$

$$= \frac{39}{850}$$

Therefore, Pr. (getting at least two diamonds) = (DDD) or (DDN) or (DND) or (NDD)

= (DDD) + (DDN) + (DND) + (NDD)

$$= \frac{11}{850} + \frac{39}{850} + \frac{39}{850} + \frac{39}{850}$$

$$= \frac{128}{850}$$

$$= \frac{64}{425}$$

(b) In order to determine the probability of getting at most one diamond, we need to compute the probabilities of the brackets that contain at most one diamond. From the outcome brackets given above, the ones that contain at most one diamond are, (DNN), (NDN), (NND), (NNN). Note that at most one, means one and below, (i.e. one and zero diamond in this case).

Hence the probability of getting at most one diamond = (DNN) + (NDN) + (NND) + (NNN)

Now, let us compute each of the probabilities. Hence:

(DNN) = Pr. (first card is a diamond) x Pr. (second card is not a diamond) x Pr. (third card is not a diamond)

$$= \frac{13}{52} \times \frac{39}{51} \times \frac{38}{50}$$

$$= \frac{1}{4} X \frac{13}{17} X \frac{19}{25}$$

$$= \frac{247}{1700}$$

(NDN) = Pr. (first card is not a diamond) x Pr. (second card is a diamond) x Pr. (third card is not a diamond)

$$= \frac{39}{52} X \frac{13}{51} X \frac{38}{50}$$

$$= \frac{3}{4} X \frac{13}{51} X \frac{19}{25}$$

$$= \frac{741}{5100}$$

$$= \frac{247}{1700}$$ (In its lowest term after equal division by 3)

(NND) = Pr. (first card is not a diamond) x Pr. (second card is not a diamond) x Pr. (third card is a diamond)

$$= \frac{39}{52} X \frac{38}{51} X \frac{13}{50}$$

$$= \frac{3}{4} X \frac{38}{51} X \frac{13}{50}$$

$$= \frac{1482}{10200}$$

$$= \frac{247}{1700}$$ (After equal division by 6)

(NNN) = Pr. (first card is not a diamond) x Pr. (second card is not a diamond) x Pr. (third card is not a diamond)

$$= \frac{39}{52} X \frac{38}{51} X \frac{37}{50}$$

$$= \frac{3}{4} X \frac{38}{51} X \frac{37}{50}$$

$$= \frac{4218}{10200}$$

$$= \frac{703}{1700}$$ (After equal division by 6)

Therefore, Pr. (getting at most one diamond) = (DNN) or (NDN) or (NND) or (NNN)

= (DNN) + (NDN) + (NND) + (NNN)

$$= \frac{247}{1700} + \frac{247}{1700} + \frac{247}{1700} + \frac{703}{1700}$$

$$= \frac{1444}{1700}$$

$$= \frac{361}{425}$$

**Practice Questions**

1. A card is picked at random from a pack of playing cards. Find the probability of picking a jack.

**Solution**

2. A card is picked at random from a pack of playing cards. Find the probability of picking a black 4.

**Solution**

3. A card is picked at random from a pack of playing cards. Find the probability of picking a red king.

**Solution**

4. A card is picked at random from a pack of playing cards. Find the probability of picking a either a black or red card.

**Solution**

5. A card is picked at random from a pack of playing cards. Find the probability of picking a black Queen.

**Solution**

6. A card is picked at random from a pack of playing cards. Find the probability of picking a card that is not an Ace.

**Solution**

7. A card is picked at random from a pack of playing cards. Find the probability of picking
(a) a queen or a king
(b) a 3 or a 9
(c) either a jack or the queen of diamonds
(d) a spade or a black 7
(e) a club or a red king
(f) a 2 or a red card

**Solution**

(a)

(c)

(d)

(e)

(f)

8. A card is picked at random from a pack of playing cards and then replaced. A second card is picked. What is the probability of picking:

(a) an 8 and a 5

(b) a black card and a 4

(c) two cards between 2 and 9 that have odd numbers

(d) two black cards

(e) two cards with the same number on them

(f) two cards with different number on them

**Solution**

(a)

(b)

(c)

(d)

(e)

(f)

9. Two cards are picked at random one after the other without replacement from a pack of playing cards. What is the probability of picking:

(a) a 4 and an ace

(b) a 2 and a 7

(c) two 8s

(d) two clubs

(e) two red cards

(f) a club and a diamond

(g) two cards that are queens

**Solution**

(a)

(b)

(c)

(d)

(e)

(f)

(g)

10. If three cards are picked from a pack of playing cards with replacement, what is the probability if getting:

(a) at least two 9s

(b) at most two 9s

**Solution**

(a)

(b)

11. If three cards are chosen from a pack of playing cards without replacement, what is the probability of getting:

(a) at least two kings

(b) at most one king?

**Solution**

(a)

(b)

Answers to Chapter 3

1. $\frac{1}{13}$ 2. $\frac{1}{26}$ 3. $\frac{1}{26}$ 4. 1 5. $\frac{1}{26}$ 6. $\frac{12}{13}$

7. (a) $\frac{2}{13}$ (b) $\frac{2}{13}$ (c) $\frac{5}{52}$ (d) $\frac{7}{26}$ (e) $\frac{15}{52}$ (f) $\frac{7}{13}$

8. (a) $\frac{1}{169}$ (b) $\frac{1}{13}$ (c) $\frac{6}{169}$ (d) $\frac{1}{4}$ (e) $\frac{1}{13}$ (f) $\frac{12}{13}$

9. (a) $\frac{8}{663}$ (b) $\frac{8}{663}$ (c) $\frac{1}{221}$ (d) $\frac{1}{17}$ (e) $\frac{25}{102}$ (f) $\frac{13}{102}$ (g) $\frac{1}{221}$

10. (a) $\frac{37}{2197}$ (b) $\frac{2196}{2197}$

11. (a) $\frac{73}{5525}$ (b) $\frac{5452}{5525}$

# CHAPTER 4
# PROBABILITY ON TOSSING OF COINS

When a coin is tossed, the outcome can either be a head or a tail. However when two or more coins are tossed, the total outcome is obtained from $2^n$, where n is the number of times the coin is tossed, or the number of coins tossed together.

Note that 'head' is the part of the coin that shows the person drawn on the coin, while the opposite side of the coin is called the 'tail'

**Examples**

1. A fair coin is tossed. What is the probability of getting:
(a) a head
(b) a tail

**Solution**

(a) There are only two possible outcomes. Head or tail.

Therefore, Pr. (getting a head) = $\frac{\text{Number of heads}}{\text{Total outcomes}}$

$= \frac{1}{2}$

(b) Pr. (getting a tail) = $\frac{\text{Number of tails}}{\text{Total outcomes}}$

$= \frac{1}{2}$

2. A coin is tossed two times. What is the probability of getting:
(a) a head and a tail
(b) at least a tail
(c) two heads
(d) two tails
(e) a head on the first toss, and a tail on the second toss.

**Solution**

The outcomes are written by using H for head and T for tail. The total number of outcomes will be $2^2$ = 4 (i.e. from $2^n$, and n = 2 in this case)

The outcomes are: (HH), (HT), (TH), (TT).

(a) The outcomes with head and tail are (HT) and (TH). This gives 2 outcomes.

Therefore, Pr. (getting a head and tail) = $\frac{\text{Number of outcome with head and tail}}{\text{Total outcomes}}$

$= \frac{2}{4}$

$= \frac{1}{2}$

(b) The outcomes with at least a tail are (HT), (TH) and (TT). This gives 3 outcomes.

Therefore, Pr. (getting at least a tail) = $\frac{\text{Number of outcomes with at least a tail}}{\text{Total number of outcomes}}$

$= \frac{3}{4}$

(c) The outcome with two heads is (HH). This gives 1 outcome.

Therefore, Pr. (getting two heads) = $\frac{\text{Number of outcomes with heads}}{\text{Total number of outcomes}}$

$= \frac{1}{4}$

(d) The outcome with two tails is (TT). This gives 1 outcome.

Therefore, Pr. (getting two tails) = $\frac{\text{Number of outcomes with two tails}}{\text{Total number of outcomes}}$

$= \frac{1}{4}$

(e) The outcome with a head on the first toss, and a tail on the second toss is (HT). This gives 1 outcome

Therefore, Pr. (getting a head on the first toss, and a tail on the second toss) = $\frac{1}{4}$

3. A coin is tossed three times. What is the probability of getting:
(a) two heads and one tail
(b) at least one head
(c) three tails

(d) at least two heads
(e) a tail, a head and a tail

**Solution**

(a) The total number of outcomes will be $2^3 = 8$ (i.e. from $2^n$, and n = 3 in this case)

The outcomes are: (HHH), (HTH), (HTT), (HHT), (THH), (THT), (TTH), (TTT). This gives a total of 8 outcomes

The outcomes with two heads and one tail are (HTH), (HHT) and (THH). This gives 3 outcomes.

Therefore, Pr. (getting two heads and one tail) = $\frac{\text{Number of outcomes with two heads and one tail}}{\text{Total number of outcomes}}$

$= \frac{3}{8}$

(b) The outcomes with at least one head are (HHH), (HTH), (HTT), (HHT), (THH), (THT) and (TTH). This gives 7 outcomes.

Therefore, Pr. (getting at least one head) = $\frac{\text{Number of outcomes with at least one head}}{\text{Total number of outcomes}}$

$= \frac{7}{8}$

(c) The outcome with three tails is (TTT). This gives 1 outcome.

Therefore, Pr. (getting three tails) = $\frac{1}{8}$

(d) The outcomes with at least two heads are (HHH), (HTH), (HHT) and (THH). This gives 4 outcomes.

Therefore, Pr. (getting at least two heads) = $\frac{\text{Number of outcomes with at least two heads}}{\text{Total number of outcomes}}$

$= \frac{4}{8}$

$= \frac{1}{2}$

(e) The outcome with a tail, a head and a tail is (THT). This is 1 outcome

Hence, Pr. (getting a tail, a head and a tail) = $\frac{1}{8}$

4. Four coins are tossed together. Find the probability of getting:

(a) two heads and two tails

(b) four tails

(c) at least three heads

(d) at least two heads

(e) one head

**Solution**

The total number of outcomes will be $2^4 = 16$ (i.e. from $2^n$, and $n = 4$ in this case)

The outcomes are: (HHHH), (HHHT), (HHTT), (HTTT), (THHH), (TTHH), (TTTH), (THTH), (HTHT), (HHTH), (THHT), HTTH), (TTHT), (THTT), (HTHH), (TTTT). This gives a total of 16 outcomes.

(a) The outcomes with two heads and two tails are (HHTT), (TTHH), (THTH), (HTHT), (THHT), (HTTH). This gives 6 outcomes.

Therefore, Pr. (getting two heads and two tails) = $\frac{\text{Number of outcomes with two heads and two tails}}{\text{Total number of outcomes}}$

$= \frac{6}{16}$

$= \frac{3}{8}$

(b) The outcome with four tails is (TTTT). This gives 1 outcomes.

Therefore, Pr. (getting four tails) = $\frac{1}{16}$

(c) The outcomes with at least three heads are (HHHH), (HHHT), (THHH), (HHTH), (HTHH). This gives 5 outcomes.

Therefore, Pr. (getting at least three heads) = $\frac{\text{Number of outcomes with at least three heads}}{\text{Total number of outcomes}}$

$= \frac{5}{16}$

(d) The outcomes with at least two heads are (HHHH), (HHHT), (HHTT), (THHH), (TTHH), (THTH), (HTHT) (HHTH), (THHT), (HTTH), (HTHH). This gives 11 outcomes.

Therefore, Pr. (getting at least two heads) = $\frac{\text{Number of outcomes with at least two heads}}{\text{Total number of outcomes}}$

$= \frac{11}{16}$

(e) The outcomes with one head are, (HTTT), (TTTH), (TTHT), (THTT), . This gives 4 outcomes.

Therefore, Pr. (getting one head) = $\frac{\text{Number of outcomes with one head}}{\text{Total number of outcomes}}$

$= \frac{4}{16}$

$= \frac{1}{4}$

(5) A coin is tossed five times. Find the probability of getting at least one tail.

**Solution**

The total number of outcomes will be $2^5 = 32$

The only outcome without a tail is (HHHHH). This is an outcome of 1

Pr. (getting no tail, i.e. all head) = $\frac{1}{32}$

Therefore, Pr. (getting at least one tail) = 1 - Pr. (getting no tail)

$= 1 - \frac{1}{32}$

$= \frac{31}{32}$

**Practice Questions**

1. A fair coin is tossed. What is the probability of getting:
(a) a tail
(b) a head
(c) a tail or a head

**Solution**

(a)

(b)

2. A coin is tossed two times. What is the probability of getting:
(a) a tail and then a head
(b) at least a head
(c) two tails
(d) at least a tail
(e) a head on the first toss, and a tail on the second toss.

**Solution**

(a)

(b)

(c)

(d)

(e)

3. Three coins are tossed. What is the probability of getting:
(a) three heads
(b) at least one tail
(c) a head, a tail and then a head
(d) at least one head
(e) at least two heads
(f) at most two tails

**Solution**

(a)

(b)

(c)

(d)

(e)

(f)

4. Four coins are tossed together. Find the probability of getting:
(a) at least one head
(b) four heads
(c) at least two heads
(d) at most three tails
(e) two heads

**Solution**

(a)

(b)

(c)

(d)

(e)

(5) A coin is tossed five times. Find the probability of getting at least one head.

## Answers to Chapter 4

1. (a) $\frac{1}{2}$ (b) $\frac{1}{2}$ (c) 1

2. (a) $\frac{1}{4}$ (b) $\frac{3}{4}$ (c) $\frac{1}{4}$ (d) $\frac{3}{4}$ (e) $\frac{1}{4}$

3. (a) $\frac{1}{8}$ (b) $\frac{7}{8}$ (c) $\frac{1}{8}$ (d) $\frac{7}{8}$ (e) $\frac{1}{2}$ (f) $\frac{7}{8}$

4. (a) $\frac{15}{16}$ (b) $\frac{1}{16}$ (c) $\frac{11}{16}$ (d) $\frac{15}{16}$ (e) $\frac{3}{8}$

5. $\frac{31}{32}$

# CHAPTER 5
# PROBABILITY ON THROWING OF DICE

**Examples**

1. A fair die is rolled once. What is the probability of getting:
(a) a number divisible by 3
(b) a multiple of 2
(c) at least 5
(d) at most 2
(e) a prime number or an even number
(f) either a number greater that 2 or a multiple of 4

**Solution**

(a) Pr. (getting a number divisible by 3) = $\frac{\text{Number of faces having numbers divisible by 3}}{\text{total number of faces}}$

$= \frac{2}{6}$ (Faces with numbers divisible by 3 are 3 and 6, i.e. 2 faces)

$= \frac{1}{3}$

(b) Pr. (getting a multiple of 2) = $\frac{\text{Number of faces having numbers that are multiple of 2}}{\text{total number of faces}}$

$= \frac{3}{6}$ (Faces with numbers that are multiple of 2 are 2, 4 and 6, i.e. 3 faces)

$= \frac{1}{2}$

(c) Pr. (getting at least 5) = $\frac{\text{Number of faces having numbers that are at least 5}}{\text{Total number of faces}}$

$= \frac{2}{6}$ (Faces with numbers that are at least 5 are 5 and 6, i.e. 2 faces)

$= \frac{1}{3}$

(d) Pr. (getting at most 2) = $\frac{\text{Number of faces having numbers that are at least 2}}{\text{Total number of faces}}$

$= \frac{2}{6}$ (Faces with numbers that are at most 2 are 1 and 2, i.e. 2 faces)

$= \frac{1}{3}$

(e) Pr. (getting a prime number or an even number) = $\frac{\text{Numbe r of faces having prime numbers and even numbers}}{\text{Total number of faces}}$

$= \frac{5}{6}$ (Faces with prime numbers are 2, 3, and 5. Faces with even numbers are 2, 4, 6. This will give a total of 5 faces because 2 which is both a prime and even number should be counted once)

Therefore, Pr. (getting a prime number or an even number) = $\frac{5}{6}$

(f) Pr. (getting either a number greater that 2 or a multiple of 4) = $\frac{\text{Number of faces having numbers greater than 2 and numbers that are multiple of 4}}{\text{Total number of faces}}$

$= \frac{4}{6}$ (Faces with numbers greater than 2 are 3, 4, 5 and 6. Faces with multiple of 4 is 4. This will give a total of 4 faces because 4 which appear in both events should be counted once)

Therefore, Pr. (getting either a number greater that 2 or a multiple of 4) = $\frac{4}{6}$

$= \frac{2}{3}$

2. A die is thrown and a coin is tossed. What is the probability of getting:
(a) a 3 and a head
(b) a tail and a prime number

**Solution**

(a) From the die, Pr. (getting a 3) = $\frac{1}{6}$

From the coin, Pr. (getting a head) = $\frac{1}{2}$

Since AND means multiplication in probability:

Therefore, Pr. (getting a 3 and a head) = Pr. (getting a 3) x Pr. (getting a head)

$= \frac{1}{6} \times \frac{1}{2}$

$= \frac{1}{12}$

(b) (a) From the coin, Pr. (getting a tail) = $\frac{1}{2}$

From the die, Pr. (getting a prime number) = $\frac{3}{6}$ (The prime numbers are 3, i.e. 2, 3 and 5)

$= \frac{1}{2}$

Therefore, Pr. (getting a tail and a prime number) = Pr. (getting a tail) x Pr. (getting a prime number)

$= \frac{1}{2} \times \frac{1}{2}$

$= \frac{1}{4}$

3. Two fair dice are thrown at the same time. Find the probability of getting:
(a) at least one six
(b) a sum of at least 10
(c) a sum of at most 5
(d) a sum less than 3
(e) a total of seven
(f) a sum that is either a prime number or a multiple of 3
(g) a sum that is either divisible by 3 or a multiple of 2

**Solution**

The outcome table is as shown below. The numbers in the bracket give the outcome from the first and second die respectively. Adding the numbers in the bracket will give the respective sum that will be obtained.

| | Number on second die | | | | | | |
|---|---|---|---|---|---|---|---|
| | + | 1 | 2 | 3 | 4 | 5 | 6 |
| | 1 | (1,1) | (1,2) | (1,3) | (1,4) | (1,5) | (1,6) |
| Number on | 2 | (2,1) | (2,2) | (2,3) | (2,4) | (2,5) | (2,6) |
| first die | 3 | (3,1) | (3,2) | (3,3) | (3,4) | (3,5) | (3,6) |

| | | | | | | |
|---|---|---|---|---|---|---|
| 4 | (4,1) | (4,2) | (4,3) | (4,4) | (4,5) | (4,6) |
| 5 | (5,1) | (5,2) | (5,3) | (5,4) | (5,5) | (5,6) |
| 6 | (6,1) | (6,2) | (6,3) | (6,4) | (6,5) | (6,6) |

The outcome table above can be presented in a more direct form by adding the values in the brackets above to obtain the sum. This is as shown below. In the table below, the numbers in the brackets represent the numbers on each die. The numbers that are not in bracket are the outcomes from the sum of numbers on first and second dice.

| | | Number on second die | | | | | |
|---|---|---|---|---|---|---|---|
| | + | (1) | (2) | (3) | (4) | (5) | (6) |
| | (1) | 2 | 3 | 4 | 5 | 6 | 7 |
| Number on | (2) | 3 | 4 | 5 | 6 | 7 | 8 |
| first die | (3) | 4 | 5 | 6 | 7 | 8 | 9 |
| | (4) | 5 | 6 | 7 | 8 | 9 | 10 |
| | (5) | 6 | 7 | 8 | 9 | 10 | 11 |
| | (6) | 7 | 8 | 9 | 10 | 11 | 12 |

Note that any of the tables above can be used to answer the questions asked above.

(a) The outcomes that can be obtained from getting at least a six are (6,1), (6,2), (6,3), (6,4), (6,5), (6,6), (1,6), (2,6), (3,6), (4,6), (5,6). They are from the first table. They are the outcomes from the 6 on the first die, and 6 on the second die respectively. The number of brackets from this outcome is 11 (when the brackets are counted). Note that the total outcomes from any of the two outcome tables above is 36. This is easier obtained from the second table by counting the numbers that are not in bracket.

Therefore, Pr. (getting at least a six) = $\frac{\text{Number of outcomes obtained when at least a six shows}}{\text{Total number of outcomes on the table}}$

$= \frac{11}{36}$

(b) A sum of at least 10 as shown on the second table above are, 10, 10, 10, 11, 11, and 12. This gives a total of 6 outcomes.

Therefore, Pr. (getting a sum of at least 10) = $\frac{6}{36}$ (Note that 36 is the total outcome)

$= \frac{1}{6}$

(c) A sum of at most 5 as shown on the second table above are, 5, 5, 5, 5, 4, 4, 4, 3, 3, and 2. This gives a total of 10 outcomes.

Therefore, Pr. (getting a sum of at most 5) = $\frac{10}{36}$

$= \frac{5}{18}$

(d) A sum less than 3 as shown on the second table above 2 only. This gives a total of 1 outcome.

Therefore, Pr. (getting a sum less than 3) = $\frac{1}{36}$

(e) A total of 7 as shown on the second table above appears 6 times. This gives a total of 6 outcomes.

Therefore, Pr. (getting a total of 7) = $\frac{6}{36}$

$= \frac{1}{6}$

(f) Sums which are prime numbers are 2, 3, 5, 7 and 11. Sums which are multiple of 3 are 3, 6, 9 and 12. Hence we are to count the outcomes from 2, 3, 5, 6, 7, 9, 11, and 12 (3 should be counted once). Hence, from the table, 2 appears 1 time, 3 appears 2 times, 5 appears 4 times, 6 appears 5 times, 7 appears 6 times, 9 appears 4 times, 11 appears 2 times, 12 appears 1 time. This gives a total outcome of 1 time + 2 times + 4 times + 5 times + 6 times + 4 times + 2 times + 1 time = 25. This is easier done on the table by counting all 2, 3, 5, 6, 7, 9, 11 and 12. It will also give a total of 25 outcomes.

Therefore, Pr. (getting a sum that is either a prime number or a multiple of 3) = $\frac{25}{36}$

(g) Sums which are divisible by 3 are 3, 6, 9 and 12. Sums which are multiples of 2 are 2, 4, 6, 8, 10 and 12. Hence we are to count the outcomes from 2, 3, 4, 6, 8, 9, 10 and 12 (6 and 12 which appear in both events should be counted once each). Hence, we go to the second table above and count all 2, 3, 4, 6, 8, 9, 10 and 12. It will give a total of 24 outcomes.

Therefore, Pr. (getting a sum that is either divisible by 3 or a multiple of 2) = $\frac{24}{36}$

$= \frac{2}{3}$

4. An unbiased die with faces numbered 1 to 6 is rolled twice. Find the probability that the product of the numbers obtained is:
(a) odd
(b) even
(c) 12
(d) prime
(e) either odd or a multiple of 5

**Solution**

The outcome table is as shown below. The numbers in brackets are the numbers on the die.

| | | Number on second die | | | | | |
|---|---|---|---|---|---|---|---|
| | x | (1) | (2) | (3) | (4) | (5) | (6) |
| | (1) | 1 | 2 | 3 | 4 | 5 | 6 |
| Number on | (2) | 2 | 4 | 6 | 8 | 10 | 12 |
| first die | (3) | 3 | 6 | 9 | 12 | 15 | 18 |
| | (4) | 4 | 8 | 12 | 16 | 20 | 24 |
| | (5) | 5 | 10 | 15 | 20 | 25 | 30 |
| | (6) | 6 | 12 | 18 | 24 | 30 | 36 |

(a) All the odd numbers from the outcome table above are 1, 3, 3, 5, 5, 9, 15, 15, 25. This gives a total of 9 outcomes.

Therefore, Pr. (product of numbers is odd) = $\frac{9}{36}$ (Note that the total outcomes is 36)

$= \frac{1}{4}$

(b) Pr. (product of numbers is even) = 1 - Pr. (product of numbers is odd)

$= 1 - \frac{1}{4}$

$$= \frac{3}{4}$$

This can also be obtained by counting all the outcomes that are even numbers in the table above. Total even numbers is 27.

Hence, Pr. (product of numbers is even) = $\frac{27}{36}$

$$= \frac{3}{4} \quad \text{(As obtained before)}$$

(c) Pr. (product of numbers is 12) = $\frac{4}{36}$ (12 appears 4 times in the table)

$$= \frac{1}{9}$$

(d) All the prime numbers are, 2, 2, 3, 3, 5, 5. This gives a total outcomes of 6.

Therefore, Pr. (product of numbers is prime) = $\frac{6}{36}$

$$= \frac{1}{6}$$

(e) All products that are odd numbers are, 1, 3, 3, 5, 5, 9, 15, 15, 25. All products that are multiples of 5 are, 5, 5, 10, 10, 15, 15, 20, 20, 25, 30, 30.

They will both give a total outcome of 15. Note that, 5, 5, 15, 15, 25 are counted once under odd number. They should not be counted under multiples of 5, as this will result to double counting. Hence with this total outcome of 15,

Pr. (product of numbers is either odd or a multiple of 5) = $\frac{15}{36}$

$$= \frac{5}{12}$$

5. Three dice are thrown together. What is the probability of getting a total score of 10?

**Solution.**

If a die is thrown once, the total outcome is given by $6^1 = 6$. If two dice are thrown, the total outcome is $6^2 = 36$. Similarly, if three dice are thrown, the total outcome will be $6^3 = 216$.

Now, for us to draw a table with 216 outcomes will be very tedious. So, a direct way of solving this problem will be to select the outcomes from each die that will result in a total score of 10. These outcomes are:

(6, 3, 1), (6, 2, 2), (5, 4, 1), (5, 3, 2), (4, 4, 2), (4, 3, 3)

Each of the brackets above can give us 6 outcomes. For example, the first bracket above can give us the following 6 outcomes:

(6, 3, 1): which means - First die shows 6, second die shows 3, third die shows 1

(6, 1, 3): which means - First die shows 6, second die shows 1, third die shows 3

(1, 6, 3): which means - First die shows 1, second die shows 6, third die shows 3

(1, 3, 6): which means - First die shows 1, second die shows 3, third die shows 6

(3, 1, 6): which means - First die shows 3, second die shows 1, third die shows 6

(3, 6, 1): which means - First die shows 3, second die shows 6, third die shows 1

Similarly, each of the other brackets can give us 6 outcomes.

Let us write out our outcome brackets again. They are, (6, 3, 1), (6, 2, 2), (5, 4, 1), (5, 3, 2), (4, 4, 2), (4, 3, 3)

When each of these brackets give us 6 outcomes, then we will obtain a total of 36 (i.e. 6 x 6) outcomes. Recall that our overall outcome table will give us a total of 216 (i.e. $6^3$) outcomes.

Therefore, Pr. (getting a total score of 10) = $\frac{36}{216}$

$= \frac{1}{6}$

**Practice Questions**

1. A fair die is thrown once. Find the probability of getting:
(a) a 5
(b) a 1
(c) a 9
(d) a 2 or 3 or 6

(e) a number less than 6
(f) a prime or an even number

**Solution**

(a)

(b)

(c)

(d)

2. A fair die is rolled once. What is the probability of getting:
(a) a number divisible by 2
(b) a multiple of 3
(c) at least 2
(d) at most 3
(e) a perfect square or an odd number
(f) either a number greater that 5 or a multiple of 3

**Solution**

(a)

(b)

(c)

(d)

(e)

(f)

3. A die is thrown and a coin is tossed. What is the probability of getting:
(a) a 5 and a head
(b) a tail and a perfect cube

**Solution**

(a)

(b)

4. Two fair dice are thrown at the same time. Find the probability of getting:
(a) at least one four
(b) a sum of at least 6
(c) a sum of at most 10
(d) a sum less than 8
(e) a total of 12

(f) a sum that is either a perfect square or a multiple of 5
(g) a sum that is either divisible by 6 or a multiple of 4

**Solution**

(a)

(b)

(c)

(d)

(e)

(f)

(g)

5. An unbiased die with faces numbered 1 to 6 is rolled twice. Find the probability that the product of the numbers obtained is:

(a) prime

(b) divisible by 6

(c) 9

(d) a factor of 10

(e) either perfect cube or a multiple of 8

**Solution**

(a)

(b)

(c)

(d)

(e)

(d)

(e)

6. Three dice are thrown together. What is the probability of getting a total score of 11?

## Answers to Chapter 5

1. (a) $\frac{1}{6}$ (b) $\frac{1}{6}$ (c) $\frac{1}{6}$ (d) $\frac{1}{2}$ (e) $\frac{5}{6}$ (f) $\frac{5}{6}$

2. (a) $\frac{1}{2}$ (b) $\frac{1}{2}$ (c) $\frac{5}{6}$ (d) $\frac{1}{2}$ (e) $\frac{2}{3}$ (f) $\frac{1}{3}$

3. (a) $\frac{1}{12}$ (b) $\frac{1}{12}$

4. (a) $\frac{11}{36}$ (b) $\frac{13}{18}$ (c) $\frac{11}{12}$ (d) $\frac{7}{12}$ (e) $\frac{1}{36}$ (f) $\frac{7}{18}$ (g) $\frac{7}{18}$

5. (a) $\frac{1}{6}$ (b) $\frac{5}{12}$ (c) $\frac{1}{36}$ (d) $\frac{1}{6}$ (e) $\frac{1}{6}$

6. $\frac{1}{6}$

# CHAPTER 6
# MISCELLANEOUS PROBLEMS ON PROBABILITY

**Examples**

1. A box contains two green balls, three yellow balls and four white balls. A ball is picked at random from the box. What is the probability that it is:

(a) green
(b) yellow
(c) white
(d) blue
(e) not white
(f) either yellow or green

**Solution**

Total number of balls in the box = 2 + 3 + 4 = 9

(a) Pr. (that it is green) = $\frac{\text{Number of green balls}}{\text{Total number of balls in the box}}$

$= \frac{2}{9}$

(b) Pr. (that it is yellow) = $\frac{\text{Number of yellow balls}}{\text{Total number of balls in the box}}$

$= \frac{3}{9}$

$= \frac{1}{3}$

(c) Pr. (that it is white) = $\frac{\text{Number of white balls}}{\text{Total number of balls in the box}}$

$= \frac{4}{9}$

(d) There is no blue ball in the box.

Therefore, Pr. (that it is blue) = 0

(e) Pr. (that it is not white) = 1 - Pr. (that it is white)

$$= 1 - \frac{4}{9}$$

$$= \frac{5}{9}$$

(f) Pr. (that it is either yellow or green) = $\dfrac{\text{Number of yellow and green balls}}{\text{Total number of balls in the box}}$

$$= \frac{3+2}{9}$$

$$= \frac{5}{9}$$

Or,

Pr. (that it is either yellow or green) = Pr. (that it is yellow) + Pr. (that it is green) (Since OR means addition)

$$= \frac{1}{3} + \frac{2}{9}$$

$$= \frac{3+2}{9}$$

$$= \frac{5}{9} \quad \text{(As obtained before)}$$

2. A letter is chosen at random from the word COMPUTER. What is the probability that it is:
(a) either in the word MORE or in the word CUT
(b) either in the word COPE or in the word CUTE
(c) neither in the word ROT nor in the word CUP

**Solution**
(a) The total number of letters in COMPUTER is 8 letters.

In the word MORE, the number of letters is 4, while in the word CUT, the number of letters is 3. They both give a total of 7 letters.

Therefore, Pr. (that it is either in the word MORE or in the word CUT) = $\frac{7}{8}$

(b) In the word COPE, the number of letters is 4, while in the word CUTE, the number of letters is 4. Without counting any letter twice (i.e. C and E), the two words give a total of 6 letters (i.e. C, O, P, E, U, T).

Therefore, Pr. (that it is either in the word COPE or in the word CUTE) = $\frac{6}{8}$ (The total number of letters in COMPUTER is 8 letters).

$= \frac{3}{4}$

(c) Out of the 8 letters in COMPUTER, the letters that are neither in the word ROT nor in the word CUP are letters M and E. They are 2 letters.

Therefore, Pr. (that it is neither in the word ROT nor in the word CUP) = $\frac{2}{8}$

$= \frac{1}{4}$

(3) In a college 80% of the boys and 45% of the girls can drive a car. If a boy and a girl are chosen at random, what is the probability that:
(a) both of then can drive a car |
(b) the boy cannot drive a car and the girl can drive a car
(c) neither of them can drive a car?
(d) one of them can drive a car

**Solution**
The probabilities are given in percentage. Hence the total for each probability is 100%

Therefore, Pr. (a boy can drive a car) = $\frac{80}{100}$

$= \frac{4}{5}$

Pr. (a boy cannot drive a car) = $\frac{20}{100}$ (i.e. 100 - 80 = 20)

$= \frac{1}{5}$ (Can also be obtained from $1 - \frac{4}{5}$)

Similarly, Pr. (a girl can drive a car) = $\frac{45}{100}$

$= \frac{9}{20}$ (After equal division by 5)

Pr. (a girl cannot drive a car) = $1 - \frac{9}{20})$

$= \frac{11}{20}$

(a) Therefore, Pr. (both of them can drive a car) = Pr. (a boy can drive a car) AND Pr. (a girl can drive a car)

= Pr. (a boy can drive a car) x Pr. (a girl can drive a car)

$= \frac{4}{5} \text{X} \frac{9}{20}$

$= \frac{36}{100}$

$= \frac{9}{25}$

(b) Pr. (the boy cannot drive a car and the girl can drive a car) = Pr. (a boy cannot drive a car) AND Pr. (a girl can drive a car)

= Pr. (a boy cannot drive a car) x Pr. (a girl can drive a car)

$= \frac{1}{5} \text{X} \frac{9}{20}$

$= \frac{9}{100}$

(c) Pr. (neither of them can drive a car) = Pr. (a boy cannot drive a car) AND Pr. (a girl cannot drive a car)

= Pr. (a boy cannot drive a car) x Pr. (a girl cannot drive a car)

$= \frac{1}{5} \text{X} \frac{11}{20}$

$= \frac{11}{100}$

(d) Since we do not know which of then can drive a car, then this case is logically explained as follows:

Pr. (one of them can drive a car) = either the boy can drive a car AND the girl cannot drive a car OR the girl can drive a car AND the boy cannot drive a car.

This in now calculated as follows:

Pr. (one of them can drive a car) = Pr. (the boy can drive a car) x Pr. (the girl cannot drive a car) + Pr. (the girl can drive a car) x Pr. (the boy cannot drive a car)

$$= (\frac{4}{5} \times \frac{11}{20}) + (\frac{9}{20} \times \frac{1}{5})$$

$$= \frac{11}{25} + \frac{9}{100}$$

$$= \frac{44 + 9}{100}$$

$$= \frac{53}{100}$$

4. The probability of a seed germinating is $\frac{2}{5}$. If three of the seeds are planted, what is the probability that:

(a) none will germinate
(b) at least one will germinate
(c) at least one will not germinate
(d) only one will germinate

**Solution**

This is a case of selection of three items from two possible events. We are going to write our outcomes in bracket like a tree diagram method. In order to write out the outcomes, let us use the letter G to represent germinate and letter N to represent not germinate.

Hence the outcomes are written as follows:

(GGG), (GGN), (GNG), (GNN), (NGG), (NGN), (NNG), (NNN)

(a) The probability that none will germinate is given by (NNN).

From the question, the probability that a seed germinate, G = $\frac{2}{5}$. Therefore the probability that it will not germinate, N = 1 - G = $1 - \frac{2}{5} = \frac{3}{5}$

Hence, G = $\frac{2}{5}$, N = $\frac{3}{5}$

Therefore, Pr. (that none will germinate) = (NNN)

$= \frac{3}{5} \times \frac{3}{5} \times \frac{3}{5}$

$= \frac{27}{125}$

(b) The outcomes of the probability that at least one will germinate are, (GGG), (GGN), (GNG), (GNN), (NGG), (NGN), (NNG). Hence we can compute each of the outcomes and add them together. But this will be tedious. An easier way of solving this problem is as explained below.

The difference between the outcome in question (a) and (b) is (NNN). This shows that subtracting (NNN) from the total probability will give us the outcomes in question (b). Recall that the total of any probability is 1. Therefore, 1 - (NNN) = outcomes in (b)

Hence, Pr. (that at least one will germinate) = 1 - (NNN)

$= 1 - \frac{27}{125}$ [Note that (NNN) = $\frac{27}{125}$ as calculated in question (a)]

$= \frac{108}{125}$

(c) The outcomes of the probability that at least one will not germinate are, (GGN), (GNG), (GNN), (NGG), (NGN), (NNG), (NNN). Similar to (b) above, the difference between this outcomes of this question and the overall outcomes is (GGG).

Therefore, Pr. (that at least one will not germinate) = 1 - (GGG)

Let us calculate (GGG) as follows:

Pr. [that all three will germinate, i.e. (GGG)] $= \frac{2}{5} \times \frac{2}{5} \times \frac{2}{5}$

$= \frac{8}{125}$

Therefore, Pr. (that at least one will not germinate) = 1 - (GGG)

$= 1 - \frac{8}{125}$

$= \frac{117}{125}$

(d) The outcomes of the probability that only one will germinate are, (GNN), (NGN), (NNG). Hence we will calculate each of these outcomes and add them together.

(GNN) = Pr. (that the first will germinate) x Pr. ( that the second will not germinate) x Pr. ( that the third will not germinate)

$$= \frac{2}{5} \times \frac{3}{5} \times \frac{3}{5}$$

$$= \frac{18}{125}$$

$$(NGN) = \frac{3}{5} \times \frac{2}{5} \times \frac{3}{5}$$

$$= \frac{18}{125}$$

$$(NNG) = \frac{3}{5} \times \frac{3}{5} \times \frac{2}{5}$$

$$= \frac{18}{125}$$

Therefore, Pr. (that only one will germinate) $= \frac{18}{125} + \frac{18}{125} + \frac{18}{125}$

$$= \frac{54}{125}$$

5. When children are born, they are equally likely to be boys or girls. What is the probability that in a family of four children:

(a) three are boys and one is a girl

(b) at least two are girls

(c) two are boys and two are girls

(d) the first and second born are girls

**Solution**

Since children are equally likely to be boys or girls, it means that the probability of having a boy is $\frac{1}{2}$, and the probability of having a girl is also $\frac{1}{2}$. This is similar to the case of tossing a coin (i.e. $\frac{1}{2}$ for head and $\frac{1}{2}$ for tail).

Therefore, the case of a family of four children is like when four coins are tossed. Refer to the example on tossing four coins in chapter 4.

Let us use B for boy and G for girl to write out the total outcomes of 16 (i.e. $2^4 = 16$) as shown below.

The outcomes are: (BBBB), (BBBG), (BBGG), (BGGG), (GBBB), (GGBB), (GGGB), (GBGB), (BGBG), (BBGB), (GBBG), (BGGB), (GGBG), (GBGG), (BGBB), (GGGG). This gives a total of 16 outcomes.

(a) The outcomes that the children are three boys and one girl are, (BBBG), (GBBB), (BBGB), (BGBB). This gives 4 outcomes.

Therefore, Pr. (three are boys and one is a girl) = $\frac{4}{16}$

$= \frac{1}{4}$

(b) The outcomes that the children are at least two girls are, (BBGG), (BGGG), (GGBB), (GGGB), (GBGB), (BGBG), (GBBG), (BGGB), (GGBG), (GBGG), (GGGG). This gives 11 outcomes.

Therefore, Pr. (at least two are girls) = $\frac{11}{16}$

(c) The outcomes that the children are two boys and two girls are, (BBGG), (GGBB), (GBGB), (BGBG), (GBBG), (BGGB). This gives 6 outcomes.

Therefore, Pr. (two are boys and two are girls) = $\frac{6}{16}$

$= \frac{3}{8}$

(d) The outcomes that the first and second born are girls are, (GGBB), (GGGB), (GGBG), (GGGG). This gives 4 outcomes.

Therefore, Pr. (the first and second born are girls) = $\frac{4}{16}$

$= \frac{1}{4}$

6. A bag contains three blue balls, four red balls and five white balls. Three balls are removed from the bag without replacement. What is the probability of getting:
(a) a white, blue and red balls in that order
(a) one of each colour
(c) at least two white balls

**Solution**

The total number of balls in the bag = 3 + 4 + 5 = 12

(a) A white, blue and red balls in that order means that the first is white, the second is blue and the third is red. This can be represented as (WBR).

Note that this is a case of without replacement. Hence after each ball is removed, the total number of ball remaining and the number of the particular ball removed are both reduced by one.

Therefore, Pr. (getting a white, blue and red balls, i.e. WBR) = $\frac{5}{12} \times \frac{3}{11} \times \frac{4}{10}$. (Notice how the total balls is reduced by 1 after each ball is removed from the bag.

$= \frac{60}{1320}$

$= \frac{1}{22}$ (After equal division by 60)

(b) Let B represent blue, R represent red and W represent white. Then the outcomes for getting one of each colour are given by: (BRW), (BWR), (RBW), (RWB), (WBR), (WRB).

Let us now calculate each of them.

(BRW) = Pr. (First is blue) x Pr. (Second is red) x Pr. (Third is white)

$= \frac{3}{12} \times \frac{4}{11} \times \frac{5}{10}$

$= \frac{1}{4} \times \frac{4}{11} \times \frac{1}{2}$

$= \frac{4}{88}$

$= \frac{1}{22}$

Similarly, each of the other five outcomes, i.e. (BWR), (RBW), (RWB), (WBR), (WRB), will each give us a value of $\frac{1}{22}$ when calculated. This is because each is obtained by multiplying 3 x 4 x 5, to give the numerator, and 12 x 11 x 10, to give the denominator, which simplifies to $\frac{1}{22}$.

Therefore, Pr. (getting one of each colour) = $\frac{1}{22} + \frac{1}{22} + \frac{1}{22} + \frac{1}{22} + \frac{1}{22} + \frac{1}{22}$

$= \frac{6}{22}$

$= \frac{3}{11}$

(c) Let us write out a different outcome for this problem. Since we are concerned about one colour, we are going to use W to represent white colour, and N to represent not a white colour. This will give us 8 outcomes in brackets as usual. The outcomes are:

(WWW), (WWN), (WNW), (WNN), (NWW), (NWN), (NNW), (NNN).

The outcomes representing at least two white balls are: (WWW), (WWN), (WNW), (NWW).

Number of white balls is 5. Therefore number of balls that are not white = 12 - 5 = 7, or blue + red = 3 + 4 = 7. (Blue and red ball are the balls that are not white balls).

Let us now calculate each of the outcomes above as follows:

(WWW) = Pr. (first is white) x Pr. (second is white) x Pr. (third is white)

$= \frac{5}{12} \times \frac{4}{11} \times \frac{3}{10}$ (Take note of the reduction in the white balls and total number of balls as each ball is removed from the bag)

$= \frac{60}{1320}$

$= \frac{1}{22}$

(WWN) $= \frac{5}{12} \times \frac{4}{11} \times \frac{7}{10}$ (Note that there are 7 balls that are not white)

$= \frac{140}{1320}$

$= \frac{7}{66}$

(WNW) $= \frac{5}{12} \times \frac{7}{11} \times \frac{4}{10}$

$= \frac{140}{1320}$

$= \frac{7}{66}$

(NWW) $= \frac{7}{12} \times \frac{5}{11} \times \frac{4}{10}$

$= \frac{140}{1320}$

$$= \frac{7}{66}$$

Therefore, Pr. (getting at least two white balls) = (WWW) or (WWN) or (WNW) or (NWW)

= (WWW) + (WWN) + (WNW) + (NWW)

$$= \frac{1}{22} + \frac{7}{66} + \frac{7}{66} + \frac{7}{66}$$

$$= \frac{3 + 7 + 7 + 7}{66}$$

$$= \frac{24}{66}$$

$$= \frac{4}{11}$$

7. A committee consist of 6 men and 4 women. A subcommittee made up of three members is randomly chosen from the committee members. What is the probability that:
(a) they are all men
(b) two of them are women?

**Solution**
Let us write out the outcome for this problem. Let M represent man, and W represent woman. This will give us 8 outcomes in brackets as usual. The outcomes are:

(WWW), (WWM), (WMW), (WMM), (MWW), (MWM), (MMW), (MMM).

(a) The total members in the committee are: 6 + 4 = 10.

The outcomes representing all men is (MMM)

Therefore, Pr. (they are all men, i.e. MMM) = Pr. (first is a man) x Pr. (second is a man) x Pr. (third is a man)

$= \frac{6}{10} \times \frac{5}{9} \times \frac{4}{8}$ (Notice the reduction in the number of men and people left, as each member is chosen from the committee).

$$= \frac{130}{720}$$

$$= \frac{13}{72}$$

(b) The outcomes showing that two of them are women are: (WWM), (WMW), (MWW)

Let us calculate each of them as follows:

(WWM) = Pr. (the first is a woman) x Pr. ( the second is a woman) x Pr. (the third is a man)

$$= \frac{4}{10} \times \frac{3}{9} \times \frac{6}{8}$$

$$= \frac{72}{720}$$

$$= \frac{1}{10}$$

$$(WMW) = \frac{4}{10} \times \frac{6}{9} \times \frac{3}{8}$$

$$= \frac{72}{720}$$

$$= \frac{1}{10}$$

$$(MWW) = \frac{6}{10} \times \frac{4}{9} \times \frac{3}{8}$$

$$= \frac{72}{720}$$

$$= \frac{1}{10}$$

Therefore, Pr. (two of them are women) = (WWM) or (WMW) or (MWW)

= (WWM) + (WMW) + (MWW)

$$= \frac{1}{10} + \frac{1}{10} + \frac{1}{10}$$

$$= \frac{3}{10}$$

8. A box contains seven blue pens and three red pens. Three pens are picked one after the other without replacement. Find the probability of picking:

(a) two blue pens

(b) at least two red pens

(c) at most two blue pens

**Solution**

Let B represent blue pen, and R represent red pen. The outcomes are:

(BBB), (BBR), (BRB), (BRR), (RBB), (RBR), (RRB), (RRR).

The total number of pens = 7 + 3 = 10

(a) The outcomes showing two blue pens are: (BBR), (BRB), (RBB)

Let us calculate each of them as follows:

(BBR) = Pr. (the first is a blue pen) x Pr. ( the second is a blue pen) x Pr. (the third is a red pen)

$$= \frac{7}{10} \times \frac{6}{9} \times \frac{3}{8}$$

$$= \frac{126}{720}$$

$$= \frac{7}{40} \quad \text{(In its lowest term after equal division by 18)}$$

$$(BRB) = \frac{7}{10} \times \frac{3}{9} \times \frac{6}{8}$$

$$= \frac{126}{720}$$

$$= \frac{7}{40}$$

Also, $(RBB) = \frac{7}{40}$ (Similar to the once above)

Therefore, Pr. (picking two blue pens) = $\frac{7}{40} \times \frac{7}{40} \times \frac{7}{40}$

$$= \frac{21}{40}$$

(b) The outcomes representing at least two red pens are: (RRR), (RRB), (RBR), (BRR)

Let us now calculate each of the outcomes as follows:

(RRR) = Pr. (first is a red pen) x Pr. (second is a red pen) x Pr. (third is a red pen)

$= \frac{3}{10} \text{ x } \frac{2}{9} \text{ x } \frac{1}{8}$ (Take note of the reduction in the red pens and total number of pens as each pen is picked from the box)

$= \frac{6}{720}$

$= \frac{1}{120}$

(RRB) = $\frac{3}{10} \text{ x } \frac{2}{9} \text{ x } \frac{7}{8}$

$= \frac{42}{720}$

$= \frac{7}{120}$

Hence, (RBR) = $\frac{7}{120}$ (This is similar to the one above)

And, (BRR) = $\frac{7}{120}$ (Same reason as above)

Therefore, Pr. (picking at least two red pens) = $\frac{1}{120} + \frac{7}{120} + \frac{7}{120} + \frac{7}{120}$

$= \frac{1 + 7 + 7 + 7}{120}$

$= \frac{22}{120}$

$= \frac{11}{60}$

(c) The outcomes that represent picking at most two blue pens are: (BBR), (BRB), (BRR), (RBB), (RBR), (RRB), (RRR). Note that at most two blue pens means 2, 1 or 0 blue pens.

Notice that there is only (BBB) missing from this outcome. This shows that it can be obtained by: total probability - (BBB). Which is: 1 - (BBB).

Let us calculate (BBB) as follows:

(BBB) = Pr. (first is a blue pen) x Pr. (second is a blue pen) x Pr. (third is a blue pen)

$= \frac{7}{10} \text{ x } \frac{6}{9} \text{ x } \frac{5}{8}$

$$= \frac{210}{720}$$

$$= \frac{7}{24} \quad \text{(After equal division by 30)}$$

Therefore, Pr. (picking at most two blue pens) = 1 - (BBB)

$$= 1 - \frac{7}{24}$$

$$= \frac{17}{24}$$

**Practice Questions**

1. A box contains 5 green balls, 8 yellow balls and 7 white balls. A ball is picked at random from the box. What is the probability that it is:

(a) green

(b) yellow

(c) white

(d) blue

(e) not white

(f) either yellow or green

**Solution**

(a)

(b)

(c)

(d)

(e)

(f)

2. A letter is chosen at random from the word NORMADIC. What is the probability that it is:
(a) either in the word MAD or in the word CORN
(b) either in the word NORM or in the word DAM
(c) neither in the word RID nor in the word CAN

**Solution**

(a)

(b)

(c)

(3) In a college 20% of the boys and 8% of the girls who had graduated from the college, graduated with distinction since the inception of the college. If a boy and a girl are chosen at random, what is the probability that:
(a) both of them will graduate with distinction
(b) the boy will not and the girl will graduate with distinction

(c) neither of them will graduate with distinction?
(d) one of them will graduate with distinction

**Solution**

(a)

(b)

(c)

(d)

4. The probability of a seed germinating is $\frac{1}{4}$. If three of the seeds are planted, what is the probability that:

(a) none will germinate
(b) at least one will germinate
(c) at least one will not germinate
(d) only one will germinate

**Solution**

(a)

(b)

(c)

(d)

5. When parents who are carriers of sickle cell disorder get married, they are equally likely to give birth to normal child and sick child. What is the probability that in a family of three children:

(a) two are normal and one is sick

(b) at least two are sick

(c) one is normal and two are sick

(d) the first is sick

(e) at most one is normal

**Solution**

(a)

(b)

(c)

(d)

(e)

6. A box contains six blue balls, three red balls and five white balls. Three balls are removed from the bag without replacement. What is the probability of getting:

(a) a white, blue and red balls in that order

(a) one of each colour

(c) at least two white balls

**Solution**

(a)

(b)

(c)

7. A committee consist of 4 men and 2 women. A subcommittee made up of two members is randomly chosen from the committee members. What is the probability that:

(a) they are all men

(b) one of them is a woman?

**Solution**

(a)

(b)

8. A bag contains 5 blue balls and seven red balls. Three balls are picked one after the other without replacement. Find the probability of picking:

(a) two blue balls

(b) at least two red balls

(c) at most two blue balls

**Solution**

(a)

(b)

(c)

## Answers to Chapter 6

1. (a) $\frac{1}{4}$ (b) $\frac{2}{5}$ (c) $\frac{7}{20}$ (d) 0 (e) $\frac{13}{20}$ (f) $\frac{13}{20}$

2. (a) $\frac{7}{8}$ (b) $\frac{3}{4}$ (c) $\frac{1}{4}$

3. (a) $\frac{2}{125}$ (b) $\frac{8}{125}$ (c) $\frac{92}{125}$ (d) $\frac{31}{125}$

4. (a) $\frac{27}{64}$ (b) $\frac{37}{64}$ (c) $\frac{63}{64}$ (d) $\frac{27}{64}$

5. (a) $\frac{3}{8}$ (b) $\frac{1}{2}$ (c) $\frac{3}{8}$ (d) $\frac{1}{2}$ (e) $\frac{1}{2}$

6. (a) $\frac{15}{364}$ (b) $\frac{145}{182}$ (c) $\frac{25}{91}$

7. (a) $\frac{2}{5}$ (b) $\frac{8}{15}$

8. (a) $\frac{7}{22}$ (b) $\frac{7}{11}$ (c) $\frac{21}{22}$

If you have any enquiries, suggestions or information concerning this book, please contact the author through the email below.

KINGSLEY AUGUSTINE

kingzohb2@yahoo.com

Twitter handle: @kingzohb2

www.ingramcontent.com/pod-product-compliance
Lightning Source LLC
Chambersburg PA
CBHW081451220526
45466CB00008B/2594